SACRAMENTO PUBLIC LIBRARY
828 "I" Street
Sacramento, CA 95814
08/21

ANIMALS OF THE FOREST

Lynx

by Patrick Perish

BELLWETHER MEDIA • MINNEAPOLIS, MN

Blastoff! Readers are carefully developed by literacy experts to build reading stamina and move students toward fluency by combining standards-based content with developmentally appropriate text.

Level 1 provides the most support through repetition of high-frequency words, light text, predictable sentence patterns, and strong visual support.

Level 2 offers early readers a bit more challenge through varied sentences, increased text load, and text-supportive special features.

Level 3 advances early-fluent readers toward fluency through increased text load, less reliance on photos, advancing concepts, longer sentences, and more complex special features.

★ **Blastoff! Universe**

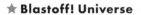

Reading Level

Grade **K**

Grades **1–3**

Grade **4**

This edition first published in 2022 by Bellwether Media, Inc.

No part of this publication may be reproduced in whole or in part without written permission of the publisher. For information regarding permission, write to Bellwether Media, Inc., Attention: Permissions Department, 6012 Blue Circle Drive, Minnetonka, MN 55343.

Library of Congress Cataloging-in-Publication Data

Names: Perish, Patrick, author.
Title: Lynx / by Patrick Perish.
Description: Minneapolis, MN : Bellwether Media, [2022] | Series: Blastoff! readers: Animals of the forest | Includes bibliographical references and index. | Audience: Ages 5-8 | Audience: Grades 2-3 | Summary: "Relevant images match informative text in this introduction to lynx. Intended for students in kindergarten through third grade"– Provided by publisher.
Identifiers: LCCN 2021004027 (print) | LCCN 2021004028 (ebook) | ISBN 9781644875070 (library binding) | ISBN 9781648344152 (ebook)
Subjects: LCSH: Lynx–Juvenile literature.
Classification: LCC QL737.C23 P3775 2022 (print) | LCC QL737.C23 (ebook) | DDC 599.75/3–dc23
LC record available at https://lccn.loc.gov/2021004027
LC ebook record available at https://lccn.loc.gov/2021004028

Editor: Kieran Downs   Designer: Josh Brink

Printed in the United States of America, North Mankato, MN.

Table of Contents

Life in the Forest

Canada lynx

Lynx are found in forests across North America, Europe, and Asia.

They have **adapted** to this often cold and snowy **biome**.

Canada Lynx Range

range =

N
W ✦ E
S

Lynx move easily through their snowy homes.

6

Their wide paws work like snowshoes. They keep the cats from sinking in deep snow.

coat

Lynx have warm fur **coats**.
The coats grow thicker in
the winter.

Dark-colored fur helps lynx
blend in with their forest homes.

tufts

Lynx have triangle-shaped ears that help them hear well.

10

Tufts on their ears may help lynx hear **rodents** under leaves or snow.

Special Adaptations

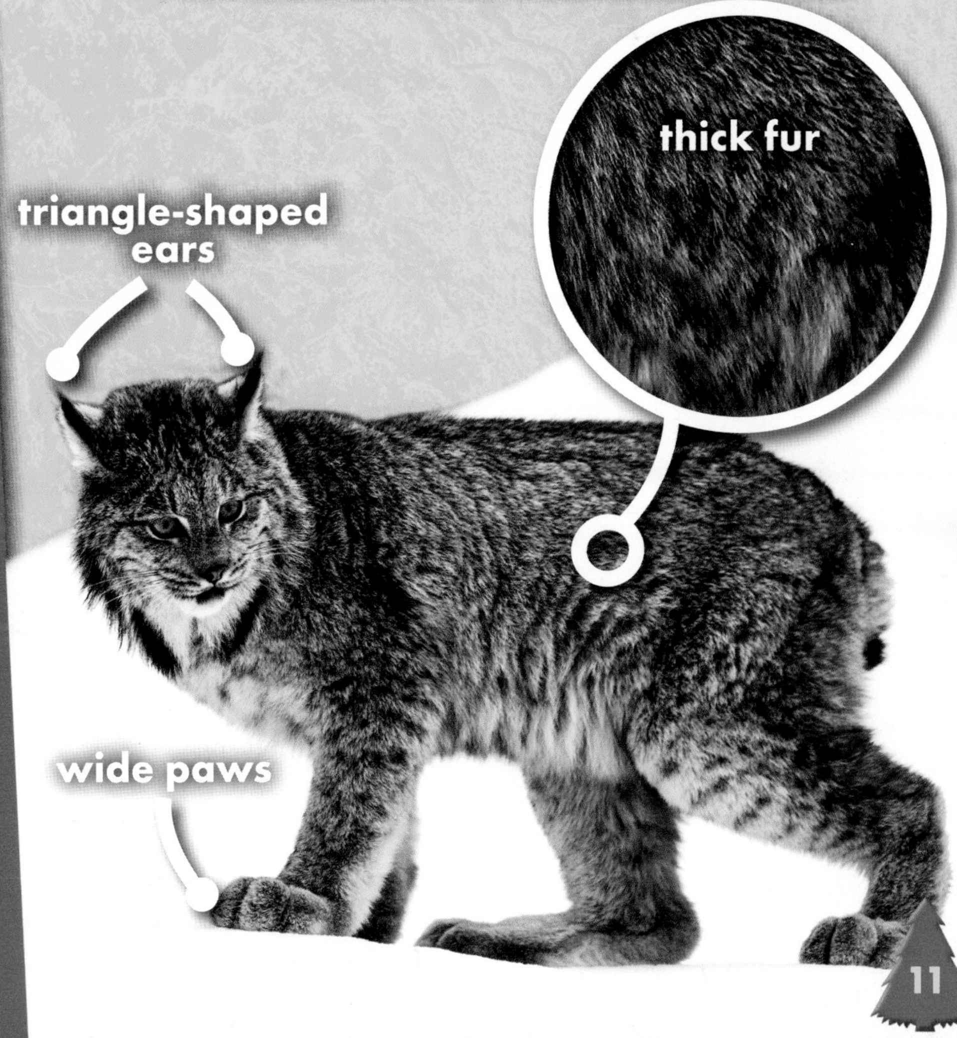

thick fur

triangle-shaped ears

wide paws

Smart Hunters

Food can be hard to find during forest winters. Lynx mark **territories** for themselves.

This helps them make sure they find enough to eat.

Eurasian Lynx Stats

Least Concern	Near Threatened	Vulnerable	Endangered	Critically Endangered	Extinct in the Wild	Extinct

conservation status: least concern

life span: up to 17 years

Eurasian lynx

Many forest animals are **nocturnal**. Lynx hunt at night so they can find **prey**.

They sleep under bushes or
fallen trees during the day.

Lynx are **carnivores**. They hunt rodents, birds, and small deer.

Lynx will also **scavenge** dead animals for food.

Lynx Diet

snowshoe hare

red squirrel

ruffed grouse

17

Lynx are **patient** hunters. They hide in trees and brush to wait for prey.

When prey finally approaches, they **pounce**!

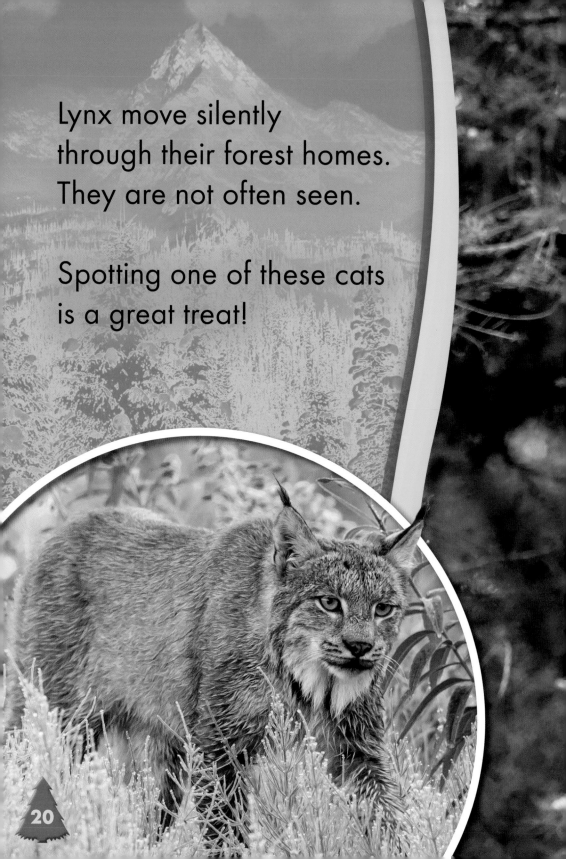

Lynx move silently
through their forest homes.
They are not often seen.

Spotting one of these cats
is a great treat!

Glossary

adapted—became well suited due to changes over a long period of time

biome—a large area with certain plants, animals, and weather

carnivores—animals that only eat meat

coats—the hair or fur covering some animals

nocturnal—active at night

patient—willing to wait a long time

pounce—to suddenly jump on something to catch it

prey—animals that are hunted by other animals for food

rodents—small animals that gnaw on their food; mice, rats, and squirrels are all rodents.

scavenge—to feed on already dead animals

territories—land areas where animals live

tufts—small bunches of hair

To Learn More

AT THE LIBRARY

Albertson, Al. *Gray Wolves.* Minneapolis, Minn.: Bellwether Media, 2020.

Carney, Elizabeth. *Wild Cats.* Washington, D.C.: National Geographic, 2017.

Thielges, Alissa. *Lynx.* Mankato, Minn.: Amicus/Amicus Ink, 2021.

ON THE WEB

FACTSURFER

Factsurfer.com gives you a safe, fun way to find more information.

1. Go to www.factsurfer.com.

2. Enter "lynx" into the search box and click 🔍.

3. Select your book cover to see a list of related content.

Index

The images in this book are reproduced through the courtesy of: Carolina K. Smith MD, cover (hero); Shairaa, cover (background); Alex Stemmer, CIP; Ghost Bear, p. 4; Danita Delimont, p. 6; Geoffrey Kuchera, p. 7; 80s_girl, p. 8; MikeLane45, p. 9; Michel VIARD, p. 10; Erickson Stock, p. 11; critterbiz, p. 11 (thick fur); Dennis W Donohue, p. 12; Dirk Rueter / Alamy, pp. 12-13; Nature Picture Library/ Alamy, p. 14; Henry Ausloos/ Alamy, p. 15; Thomas Kitchin & Vict, p. 16; FotoRequest, p. 17 (snowshoe hare); Menno Schaefer, p. 17 (red squirrel); Larry Dallaire, p. 17 (ruffed grouse); wonderful-Earth.net/ Alamy, p. 18; blickwinkel/ Kaufung/ Alamy, p. 19; Jukka Jantunen, p. 20; Evelyn D. Harrison, pp. 20-21; Milan Rybar, p. 23.